NISTIR 7478

Airtightness, Ventilation and Energy Consumption in a Manufactured House: Pre-Retrofit Results

Steven Nabinger
Andrew Persily
Building and Fire Research Laboratory

May 2008

U.S. Department of Commerce
Carlos M. Gutierrez, *Secretary*

National Institute of Standards and Technology
James Turner, Acting *Director*

ABSTRACT

A retrofit study is being conducted to investigate the airtightness, ventilation and energy impacts of tightening the exterior envelope and the heating and air conditioning system ductwork of a double section manufactured house. This report describes the results of the pre-retrofit assessment of building airtightness, ventilation, and energy consumption under both heating and cooling conditions. Measurements of building envelope airtightness and duct leakage were made using fan pressurization. Air change rates were measured continuously using the tracer gas decay technique. Energy consumption was monitored through measurement of gas consumption by the forced-air furnace for heating and electricity use by the air-conditioning system for cooling. After the pre-retrofit data collection was finished, the house underwent retrofits to tighten the building envelope and to reduce ventilation system duct leakage. After completion of the retrofits, post-retrofit data was collected. This report presents the results of the pre-retrofit measurements and a description of the retrofits.

Keywords: duct leakage, energy consumption, manufactured housing, mechanical ventilation, residential, retrofit, ventilation.

Table of Contents

1. INTRODUCTION 1
2. DESCRIPTION OF HOUSE AND VENTILATION SYSTEMS 2
3. MEASUREMENT METHODS 5
   3.1 Air Change Rates 5
   3.2 Environmental Conditions 6
   3.3 Sample locations 7
   3.4 System Airflows 8
   3.5 Building envelope and duct airtightness 9
   3.6 Energy Consumption 9
4. PRE-RETROFIT MEASUREMENT RESULTS 9
   4.1 Airtightness 9
   4.2 System flows 10
   4.3 Air Change Rates 12
   4.4 Energy Consumption 17
5. AIRFLOW MODELING ANALYSIS 19
6. DISCUSSION 25
7. ACKNOWLEDGEMENTS 27
8. REFERENCES 27

1. INTRODUCTION

Single-family residential buildings have historically been ventilated via stack and weather-driven infiltration through unintentional leakage sites in the building envelope. More recently, there has been a trend towards the use of mechanical ventilation to provide more predictable ventilation rates and air distribution that are less dependent on weather conditions. American Society of Heating, Refrigerating and Air- Conditioning Engineers (ASHRAE) Standard 62.2, Ventilation and Acceptable Indoor Air Quality in Low-Rise Residential buildings, was first published in 2003 and requires mechanical ventilation in many U.S. climates (ASHRAE 2007). However, only a very small fraction of site-built, low-rise residential buildings employ mechanical ventilation. The situation is different from manufactured homes, for which the U.S. Department of Housing and Urban Development (HUD) Manufactured Home Construction and Safety Standards (MHCSS) (HUD 1994) contain requirements for mechanical ventilation for these dwellings. In the implementation of the HUD standards, as well as the implementation of mechanical ventilation to low-rise residential buildings in general, questions have arisen regarding the actual ventilation rates in homes built to the HUD and other standards. For manufactured homes specifically, these questions concern the approaches being used to provide mechanical ventilation, and the energy and indoor air quality impacts of mechanical ventilation (Lubliner 1997 et al.). Other questions exist regarding how duct leakage, local exhaust fans and ventilation inlets affect ventilation rates, air movement patterns, and building pressures.

In order to obtain insight into these issues, a modeling study was performed on a manufactured home using the multizone airflow and indoor air quality simulation program CONTAM (Walton and Dols 2005), developed at the National Institute of Standards and Technology (NIST), to investigate different ventilation scenarios (Persily and Martin 2000). In that study, simulations were performed to predict ventilation rates due to infiltration and mechanical ventilation, interzone airflow rates, building air pressures, and ventilation air distribution. The results showed that the assumption in the HUD standard of a single value of 0.25 h^{-1} for the weather-driven infiltration rate is inherently problematic given the strong dependence of infiltration on weather. The simulated infiltration rates vary by as much as 5 to 1 based on variations in weather conditions alone. Including the impacts of exhaust fan and forced-air fan operation more than doubles the range of variation. In addition, the predicted infiltration rates were lower than this assumed value under milder weather conditions. Therefore, assuming an infiltration rate of 0.25 h^{-1} in modern manufactured homes may be too high but, more importantly, ignores variations due to weather and fan operation.

The previous modeling study also showed that employing an outdoor air intake duct on the forced-air return duct is effective in raising air change rates and distributing ventilation air throughout the house. However, the overall impact on the building air change rate is a strong function of the operating schedule of the forced-air system, which in turn depends on the extent of system oversizing and the use of other control strategies such as manual switches and timers. While increased forced-air fan operation provides higher ventilation rates, there is an energy cost associated with the increased fan operation, particularly when the forced-air fan has a high wattage rating. Also, given the existence of significant duct leakage, this approach is associated with excessive air change rates (relative to the requirements in standards) particularly when weather-driven infiltration is high. The use of whole house exhaust with passive air inlet vents also provides adequate ventilation and reasonable air distribution, but again the impact is highly dependent on the fan operation schedule. As implemented in the house model used in the simulations, the vents themselves were not particularly effective in ventilating the building since they constituted only a relatively small increase (roughly 15 %) in house leakage.

While the earlier simulation study is quite informative in terms of ventilation strategies in manufactured homes, and presumably other low-rise residential buildings, it is important to verify the findings experimentally. Therefore, in order to investigate these and other issues in the field, and to validate the conclusions of the simulation study, a double-wide manufactured house was constructed on the NIST campus. This paper presents the first results of this field study, which focused on characterizing the building's airtightness and ventilation. The paper describes the house, including the ventilation systems included to investigate different approaches to meeting the HUD requirement, the measurement equipment and techniques used in the study and the results of these measurements. In addition, a CONTAM model of the building is described and the predicted results are compared with measurements. The measurements presented were taken before the house was retrofitted to reduce envelope and duct leakage. The retrofit efforts, as well as the pre-retrofit energy consumption of the house are also described in this paper.

2. DESCRIPTION OF HOUSE AND VENTILATION SYSTEMS

This study was performed in a double-wide manufactured home installed on the NIST campus, as shown in Figure 1. Figure 2 is a schematic floor plan of the test house, showing the three bedrooms, two baths, kitchen, and the family, dining and living areas. The house has a floor area of 140 m^2 (1500 ft^2) and a volume of 340 m^3 (12,000 ft^3) and a cathedral ceiling over its full length that is 2.7 m (9.0 ft) high at the center and slopes down to 2.1 m (7.0 ft) at the front and back walls.

The exterior construction consists of fiberglass insulated 2 in x 4 in wood-frame walls, with exterior vinyl siding and vinyl covered drywall in some rooms and textured drywall in others. A vapor retarder is included in the walls, ceiling and floor. The house was installed on a block foundation over a crawl space with a floor of polyethylene sheeting under crushed stone. The crawl space block walls were treated on the outside with a liquid-applied sealant to limit moisture penetration from the surrounding soil. The crawl space has a floor area of 140 m^2 (1500 ft^2) and a volume of roughly 140 m^3 (5000 ft^3). The crawl space contains a so-called "belly" space, consisting of two sections corresponding to the front and rear of the house. The crawl space also contains eight 20 cm x 41 cm (8 in x 16 in) vents to the outdoors, 5 in the back of the house and 3 in the front. Each vent has a net free area of approximately 315 cm^2 (45 in^2), for a total free area of 0.26 m^2 (2.8 $ft.^2$) when all vents are open. During the tests described in this paper, two front vents were partially open, for a total free area of about 390 cm^2 (60 in^2). Access to the crawl space is provided by a 0.6 m x 0.6 m (2 ft x 2 ft) entry door. As seen in Figure 3, the belly contains the HVAC system ductwork, and an insulated plastic sheet separates the belly from the vented portion of the crawl space. The house has an asphalt-shingled roof with five roof vents and eave vents spanning its perimeter. The measured open area of each roof vent is 0.135 m^2 (1.45 ft^2). The eave vents have an estimated leakage area of 106 cm^2/m of eave length (16.4 in^2/ft). The attic, above the cathedral ceiling of the house, has a floor area of 142 m^2 (1,530 ft^2) and a volume of 43.9 m^3 (1,550 ft^3). The attic is insulated with 20 cm (8 in) thick paperbacked glass fiber insulation. There are a total of eleven double-hung windows installed on the north, south and west walls as shown in Figure 2. Each window contains an air vent at the top, which can be opened manually to provide additional ventilation.

Figure 1: Photograph of the manufactured house

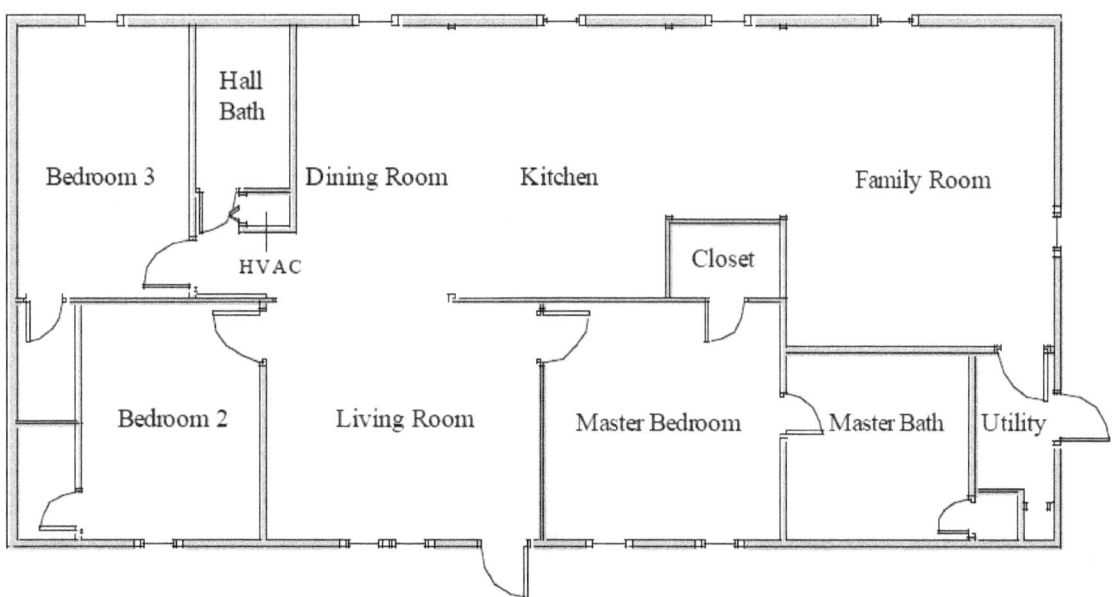

Figure 2: Schematic of floor plan of manufactured house

The test house has a heating, ventilating and air-conditioning (HVAC) system with a 22 kW (77,000 Btu/h) gas furnace, a 15 kW (3 ton) air conditioner, and a forced air fan with a design airflow rate of 470 L/s (1000 cfm). The manufacturer's specification for the air conditioner includes a coefficient of performance (COP) of 2.93. The specified efficiency of the heating system is 85 %. In addition the house contains a whole house, kitchen, and two bathroom exhaust fans. The whole house fan, located outside the hall bath, and bathroom fans each have an airflow rating of 24 L/s (50 cfm). The kitchen exhaust fan has an airflow rating of 47 L/s (100 cfm). The house is equipped with a gas cooking stove and fireplace, and an electric hot water heater. However, for these tests the fireplace was sealed off from the interior of the house with plastic sheeting.

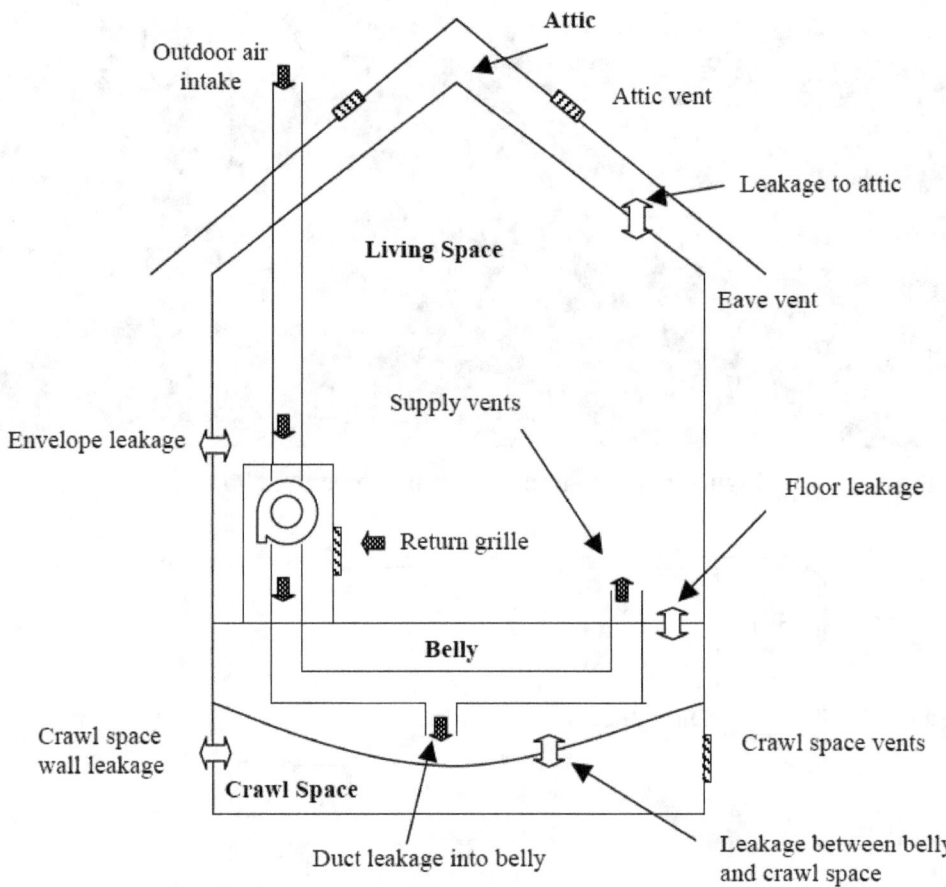

Figure 3: Schematic elevation of manufactured house

There are several strategies for ventilating the house. One option is simply to rely on envelope infiltration driven by the wind and indoor-outdoor temperature differences, as is done in most single-family dwellings. Infiltration may be supplemented by periodic operation of local exhaust fans in the bathrooms and kitchen. The house can also be ventilated via the outdoor air intake duct on the return of the forced air fan, which draws in outdoor air whenever the fan is operating. This fan can operate either continuously or as controlled by the thermostat. In addition, the whole house exhaust fan can be operated with or without the passive window vents open. The air change rate measurements described in this report were performed under these different conditions as described in section 4.

3. MEASUREMENT METHODS
This section describes the techniques and instrumentation used to make continuous measurements of building air change rates under different weather and system operating conditions using the tracer gas decay technique. The equipment and techniques used to monitor indoor and outdoor air temperatures, wind speed and direction, indoor and outdoor relative humidity, pressure differences across the building envelope, fan operation, and system airflow rates, and energy consumption are also described. A PC-based data acquisition system was installed for controlling measurement equipment and instrumentation, signal processing, data analysis and recording results.

3.1 Air Change Rates
Whole house air change rates were measured using the tracer gas decay technique as described in ASTM test method E741 (ASTM 2006). This technique determines the rate at which outdoor air enters the house, including both intentional mechanically induced outdoor air and unintentional infiltration through leaks in the building envelope. The air change rate of a building depends on a variety of factors including the design and interior configuration, installation and operation of any mechanical ventilation systems, airtightness of the envelope, and outdoor weather conditions. Therefore, many air change rate measurements, under a range of weather and system operation conditions, are required to understand the air change characteristics of a building.

Figure 4: Automated tracer gas system

In these tracer gas measurements, a tracer gas is released into the house and mixed within the interior air volume. Once the tracer gas concentration is sufficiently uniform throughout the house, the tracer gas decay concentration is monitored over time. The rate of decay of the logarithm of concentration is equal to the air change rate of the house during the measurement period in units of air changes per hour (h^{-1}). The tracer gas concentration decay technique is based on the assumption that the tracer gas concentration within the entire building can be characterized by a single value. The ASTM test method requires that the concentration within the building be uniform within 10 %, which can be achieved with tracer gas injection strategies that distribute the tracer gas throughout the building and mixing the interior air with the air handling system or with separate mixing fans.

The air change rate measurements in the house were made with an automated tracer gas system utilizing sulfur hexafluoride (SF_6) as the tracer gas. This system, shown in Figure 4, injects tracer gas into the building and then monitors the concentration at multiple locations. The tracer gas analyzer consists of a gas chromatograph (GC) equipped with an electron capture detector (ECD). The GC utilizes two 0.9 m (3 ft) long, 3 mm (1/8 in) diameter polar columns packed with molecular sieve 5a. The ECD is capable of measuring SF_6 concentrations over a range of 0.018 mg/m^3 (3 ppb(v)) to 1.8 mg/m^3 (300 ppb(v)), with an uncertainty of about 5 % of the reading. The analyzer was calibrated weekly over this range with certified SF_6 calibration gas standards.

Tracer gas was injected every 4 h to 6 h, at a rate intended to achieve an initial tracer gas concentration of about 0.7 mg/m^3 (120 ppb(v)). The tracer gas was injected into the return side of the furnace for tests with the forced-air fan on continuously and in the center of the combined kitchen, dining, family and living area for tests with the furnace fan off or in the thermostatically controlled mode. After injection, the gas was allowed to mix for a period of 10 min to obtain a uniform concentration throughout the house. The house's forced-air system was used to distribute and mix the tracer gas in many of the tests. However, in tests with the forced-air fan off or controlled by the thermostat, three mixing fans were operated. Tracer gas concentrations were then monitored at seven indoor locations, one outdoor location, the crawl space and the attic, with the concentration at each location measured once every ten minutes. In these tests each room was typically at a concentration within ±5 % of the average house concentration. The uncertainty in the whole house air change rates were typically within ±10 %.

3.2 Environmental conditions

Epoxy coated polymer thermistors were used to measure indoor and outdoor air temperatures. The indoor thermistors are capable of measuring temperatures over a range of 0 °C to 100 °C (32 °F to 212 °F) with an uncertainty no greater than 0.5 °C (32.9 °F). The outdoor thermistor is capable of measuring temperatures over a range of -30 °C to 50 °C (-22 °F to 122 °F) with an uncertainty no greater than 0.4 °C (0.7 °F). The indoor and outdoor temperatures were monitored every 100 ms, with an average value recorded every minute. The relative humidity, differential pressure, wind speed and direction, and fan switches mentioned in the following paragraphs were also monitored every 100 ms and the average of each minute recorded every minute.

Relative humidity was monitored with capacitive thin-film polymer sensors that are described in the product literature as having an uncertainty within 2 % RH of the measured value over the range of 0 % RH to 90 % RH and within 3 % RH over the range of 90 % RH to 100 % RH for temperature conditions between -40 °C and 60 °C (-40 °F and 140 °F). These sensors were calibrated over the range of 10 % to 90 % RH using a two-pressure humidity generator having an uncertainty of -0.5 % RH to +0.5 % RH. After calibration the probes were determined to be uncertain within 1 % RH over the calibrated range.

Differential pressure sensors were installed to measure the pressures across the building envelope. Two sensors were located on each exterior wall, one 10 cm (4 in) from the top and another 10 cm (4 in) from bottom of the walls. An additional sensor was installed to measure the pressure difference from the house to the attic, and two others were installed across the floor to the crawl space. These transducers are capable of measuring pressure differences of +25 Pa to -25 Pa (0.1 in wg to -0.1 in wg) with an uncertainty of 0.5 Pa (0.002 in wg) and a resolution of 0.1 Pa (0.0004 in wg).

A sonic anemometer was used to measure wind speed and direction at the top of a 10 m (33 ft) tower located approximately 5 m (16 ft) south of the southernmost wall of the house (see Figure 1). The anemometer is capable of measuring wind speeds over the range of 0 m/s to 50 m/s (0 mph to 110 mph) with an uncertainty of 0.5 m/s (1 mph) or 5 % and a resolution of 0.1 m/s (0.2 mph). It is capable of measuring wind direction within 5 % of the actual value at wind speeds greater than 4.5 m/s (10 mph). No uncertainty specification is given by the manufacturer for wind direction at lower wind speeds.

Fan status switches were wired into the electrical circuits of the forced air fan and the four exhaust fans to detect whether the fan was on or off. The switches consisted of solenoids installed in the AC power circuit of each fan. When a fan was turned on, a DC voltage signal was sent to the data acquisition system.

3.3 Sample locations

Table 1 lists the locations of tracer gas sampling tubes, temperature and relative humidity probes, fan operation switches, and differential pressure transducers. The locations are shown schematically in Figure 5. Each 'x' in the table represents a single sample location. The listing of two x's in the pressure difference column corresponds to the two pressure measurements on each of the exterior walls and the two measurements from the living space to the crawl space. The characters 's' and 'r' represent the return and supply sides of the furnace.

Table 1: Measurement locations

Location	Tracer gas	Temperature	Relative humidity	Fan operation	Pressure difference
Bedroom #3	x	x	x		x x
Main Bath	x		x	x	
Dining Room	x	x	x	x	
Kitchen	x	x		x	x x
Family Room	x	x	x		x x
Bedroom #2	x	x	x		
Living Room	x	x	x		x x
Bedroom #1	x	x	x		
Master Bath	x			x	
Furnace	s, r	s, r	s, r	x	
Attic	x x	x	x		x
Crawl Space	x x	x	x		x x
Outdoor	x x	x	x		

The floor plan in Figure 5 shows the locations of the automated monitoring system and the sample sites monitored. The types of samples are designated by the following symbols:

- ○ = tracer gas concentration
- ◉ = temperature
- ● = relative humidity
- □ = differential pressure
- ■ = fan operation

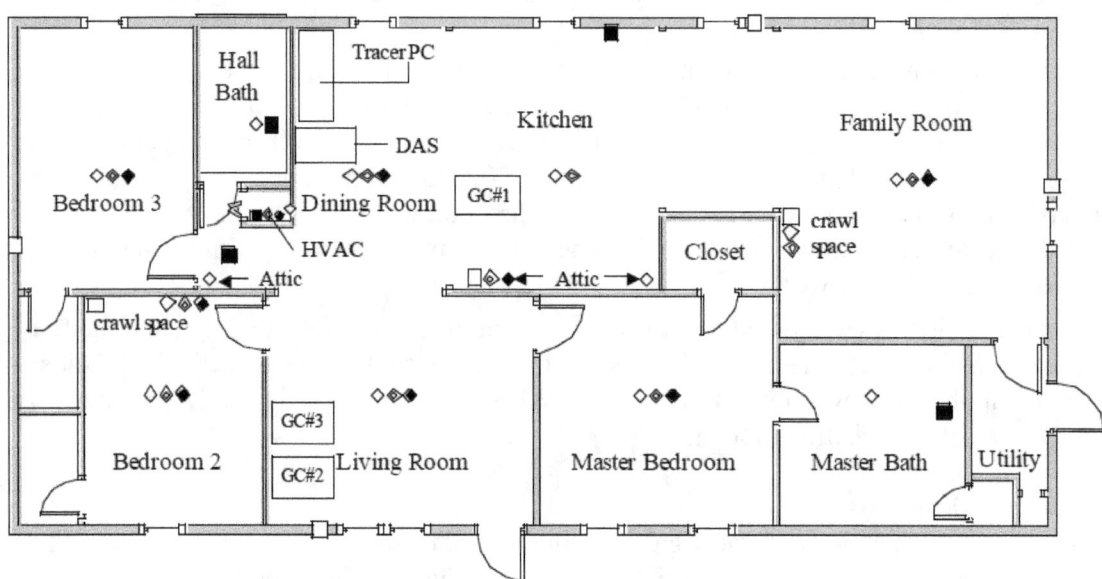

Figure 5: Sample locations of manufactured house

The indoor tracer gas sample tubes, temperature and humidity probes were positioned 1.5 m (5 ft) above the floor in the center of each room. The attic sample location is in the approximate vertical and horizontal center of the attic. The two crawl space sample locations are centered between the north and south walls, and midway between the crawl space floor and the house sub-floor (belly).

3.4 System Airflows

The airflow rates through the HVAC system and the four local exhaust fans were measured on several occasions. The HVAC system flows included the return airflow rate, the supply airflow at each vent and the outdoor air intake rate through the intake duct. A differential pressure grid was used to measure the return airflow rate into the system through the return grill, and a flow hood was used to measure the airflow rate through the outdoor air intake duct into the return side of the furnace. The airflow rates of the individual supply vents and the exhaust fans were also measured with a flow hood.

The differential pressure grid consists of a plate with a series of round metering holes and two pressure sensing grids on the upstream side (a "total pressure" grid and a "static pressure grid"). The metering plate is sealed over the furnace return grille so that air passes through the metering holes across the two grids. The airflow through the device is determined by measuring the pressure difference between the two sensing grids and using the manufacturer's calibration to convert the pressure differential to an airflow rate. The manufacturer's stated uncertainty for the device is ±7 % when used with a pressure gage of 1 % uncertainty over the range of 0.17 m^3/s to 0.99 m^3/s (365 cfm to 2,100 cfm). The uncertainty of the flow hood is stated to be 3 % +1 L/s (3 % +2 cfm) of the measured flow rate over the range of 4.7 L/s to 236 L/s (10 cfm to 500 cfm).

3.5 Building envelope and duct airtightness
The exterior envelope leakage of the house was measured by whole building pressurization testing per ASTM E779 (ASTM 2003a) using a blower door. Additional tests using two pressurization fans were used to estimate the leakage from the living space to the belly, the belly to the crawl space, and the crawl space to outside. The blower doors were calibrated using a fan calibration chamber in accordance with ASTM (2003b). The measurements of airtightness with the blower door have a relative uncertainty of roughly ±10 % of the measured value based on the calibrations performed. Pressurization tests were also used to determine the leakage from the air distribution system using ASTM E1554 (ASTM 2003c), which has an associated measurement uncertainty of about ±10 %.

3.6 Energy Consumption
Energy consumption for heating and cooling was monitored to assess the impact of air change rates on energy consumption and the energy reductions due to the airtightness retrofits. The electrical energy consumption of the air conditioning system was measured using a 240 V power transducer. The energy used by the forced-air fan was monitored with 120 V energy meters. The energy use by other items in the house, and therefore contributing to the interior heat gain, were monitored separately using additional 120 V energy meters. The 240 V power transducers have an uncertainty of ±0.05 % of full scale and a resolution of 0.1 W. The 120 V energy meters have an uncertainty of ±0.02 % of full scale and a resolution of 3.6 W·h. The heating energy was measured using a calibrated gas flow meter that recorded the natural gas flow rate into the furnace. The gas meter has a measurement range of 0 L/s to 0.94 L/s (0 cfm to 2 cfm) and an uncertainty of 1 % of full scale and a resolution of 0.0002 L/s (0.0004 cfm). All monitors (electric and gas) provided outputs every 100 ms, and average values were recorded every minute. Using the manufacturer's specifications and test data, the overall uncertainty for both the electric and gas energy measurements is less than 5 %.

4. PRE-RETROFIT MEASUREMENT RESULTS
This section contains the results of the measurements performed before the airtightening retrofits, including the house and duct airtightness, system airflows, building air change rates, and energy consumption.

4.1 Airtightness
Measurements were made to determine the exterior envelope leakage, the leakage between the living space and the crawl space, and the leakage of the ductwork. The whole building pressurization testing yielded an air change rate of 11.8 h^{-1} at 50 Pa (0.20 in wg) and an effective leakage area (ELA) of 728 cm^2 (113 in^2) at 4 Pa (0.016 in wg) with the ventilation system unsealed. Tests with the supply vents and return grill sealed yielded an air change rate of 10.1 h^{-1} at 50 Pa (0.20 in wg) and an ELA of 636 cm^2 (99 in^2) at 4 Pa (0.016 in wg). The difference between the tests with the vents sealed and unsealed is indicative of significant leakage associated with the air distribution ductwork.

A series of pressurization tests using two blower doors were conducted to determine the leakage from the living space to the belly, from the crawl space to the belly, and from the crawl space to outside. The ELA from the living space into the belly was estimated to be 510 cm^2 (80 in^2), 258 cm^2 (40 in^2) from the crawl space to the belly, and 787 cm^2 (122 in^2) from the crawl space to the outside. Due to additional leakage occurring though the crawl space walls, this latter value is about twice the net free area of the open vents (described in Section 2 of this report).

The air distribution system leakage was measured using ASTM E1554 (ASTM 2003), yielding an ELA of 320 cm^2 (50 in^2). Comparing this measurement to other recently-constructed U.S. manufactured homes reveals that the duct leakage is fairly typical, though on the high end (Persily and Martin 2000). Note that duct leakage values of roughly one-half of the measured value can be achieved through energy efficient construction techniques (Hales et al. 1997).

4.2 System flows

Airflow rates were measured for the forced air and local exhaust systems on several occasions. Table 2 summarizes the supply airflow measurements in the form of mean values for a series of measurements over several years. The variation in these flows over time was not significant. Table 2 also includes the subtotals for the front and back sections of the house and the total for all the supply vents. Figure 6 is the floor plan of the house showing the numbered supply vents from Table 2. The supply vent airflow rates range from about 15 L/s to 30 L/s (40 cfm to 60 cfm), with the exception of vent #6 in bedroom #3, which has a much higher airflow rate. This vent is close to the main supply trunk and upstream of a crushed, and therefore restricted, crossover duct, which may explain the high value. The table also shows a significant difference in the supply airflow between the front and rear sections of the house. This difference may also be due to the partially crushed crossover duct, which is close to the main fan and runs between vent #6 in the rear and vent #7 in the front sections of the house.

Table 2: Supply airflow rate measurements

Room	Vent No.	Average Airflow, L/s (cfm)
Family	1	39 (83)
Family	2	32 (67)
Kitchen	3	33 (69)
Dining	4	32 (67)
Bath2	5	22 (47)
Bed3	6	67 (142)
Bed2	7	15 (32)
Living	8	18 (39)
Living	9	21 (44)
Bed1	10	22 (47)
Bed1	11	16 (34)
Bath1	12	12 (25)
Utility	13	15 (32)
Subtotal-back	1 to 6	224 (475)
Subtotal-front	7 to 13	120 (255)
Total	**1 to 13**	**344 (729)**

Table 3 summarizes the airflow rate measurements of the four exhaust fans, the system outdoor air intake duct and the return grill as mean values for a series of measurements over several years. The mean return airflow rate is 478 L/s (1013 cfm), which is 134 L/s (284 cfm) greater than the total airflow, 344 L/s (729 cfm), through the supply air vents. This difference is presumably due to duct leakage in the air distribution system.

The outdoor intake and exhaust airflow measurements are also presented in Table 3. The mean airflow rate into the outdoor air intake is 8 L/s (16 cfm), which is about one-third of the ventilation requirement based the HUD MHCSS of 25 L/s (53 cfm). That standard requires a mechanical ventilation rate of 0.18 L/s per m^2 of floor area (0.035 cfm/ft^2). The mean measured airflow rates of the bathroom and whole house exhaust fans range from 7 L/s to 8 L/s (15 cfm to 16 cfm), which is well below their rated airflow of 24 L/s (50 cfm). The mean measured airflow rate of the kitchen exhaust is 27 L/s (57 cfm), which is only about half the its rated capacity of 47 L/s (100 cfm). Note that the rated flows are consistent with the HUD requirements.

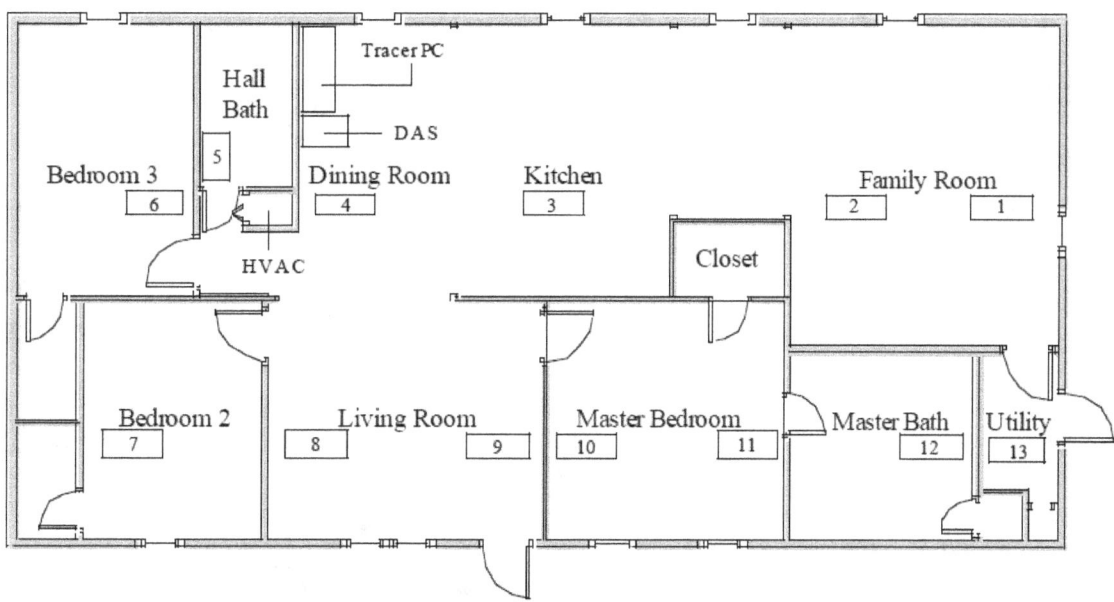

Figure 6: HVAC supply vent locations

Table 3: Exhaust, outdoor air intake and return airflow rate measurements

	Average airflow rate, L/s (cfm)
Exhaust fans	
Bath 1	7 (15)
Bath 2	8 (16)
Whole House	7 (15)
Kitchen	27 (57)
Outdoor air intake	8 (16)
Return grille	478 (1013)

4.3 Air Change Rates

Air change rates were measured with the tracer gas decay technique under the five house/system configurations listed in Table 4. During these tests all interior doors were open, all windows closed, all passive window vents closed, all exhaust fans off, and the fireplace sealed off from the interior space.

Table 4: Air change rate test conditions

Test Condition	Description
0	All fans off, outdoor air intake sealed, infiltration only
1a	Forced air fan on continuously, outdoor air intake sealed
1b	Forced air fan on continuously, outdoor air intake open
2a	Forced air fan controlled by thermostat, outdoor air intake sealed
2b	Forced air fan controlled by thermostat, outdoor air intake open

Figure 7 is a plot of the air change rates versus indoor-outdoor temperature difference for condition 0, i.e., infiltration only. This figure differentiates between data at low and high wind speeds. An approximately linear relationship is seen between the whole house air change rate and temperature difference at low wind speeds (i.e. < 2 m/s (4.5 mph)).

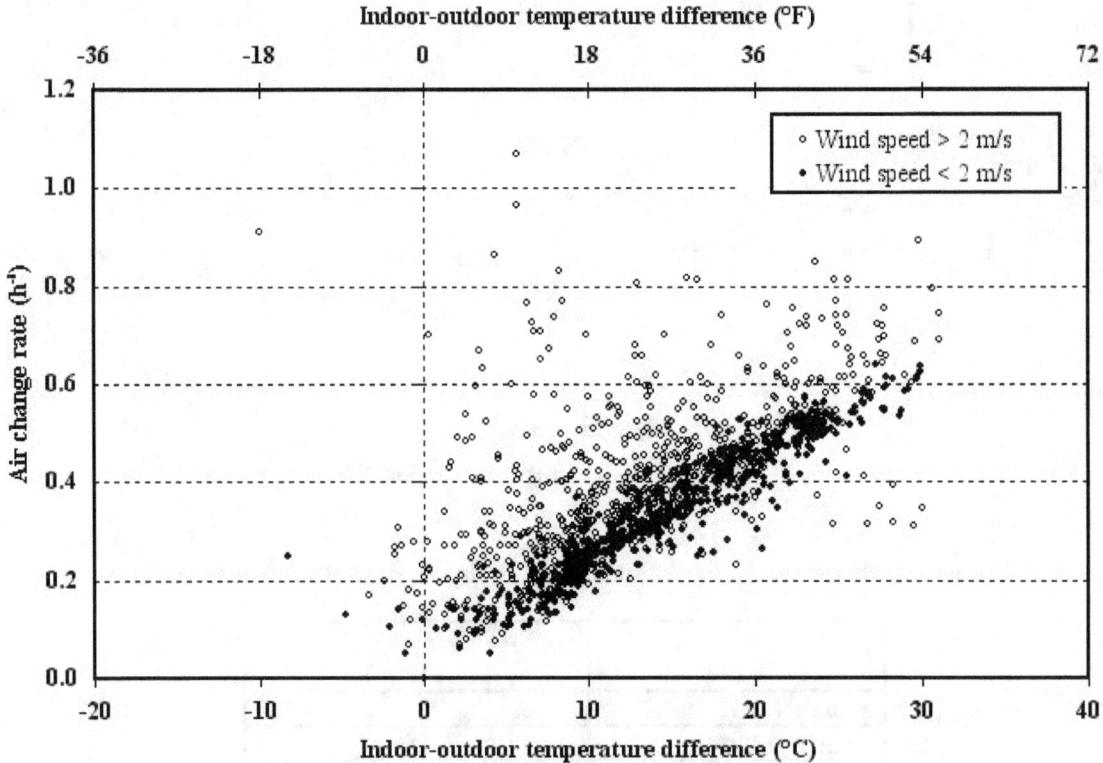

Figure 7: Air change rate versus indoor-outdoor temperature difference, Condition 0

Figure 8 shows the dependence of air change rate on wind speed for Condition 0, with the data restricted to indoor-outdoor temperature differences between -10 °C and +10 °C (-18 °F and +18 °F). The relationship is fairly linear for wind speeds greater than 3 m/s (6.7 mph), but there is still significant scatter.

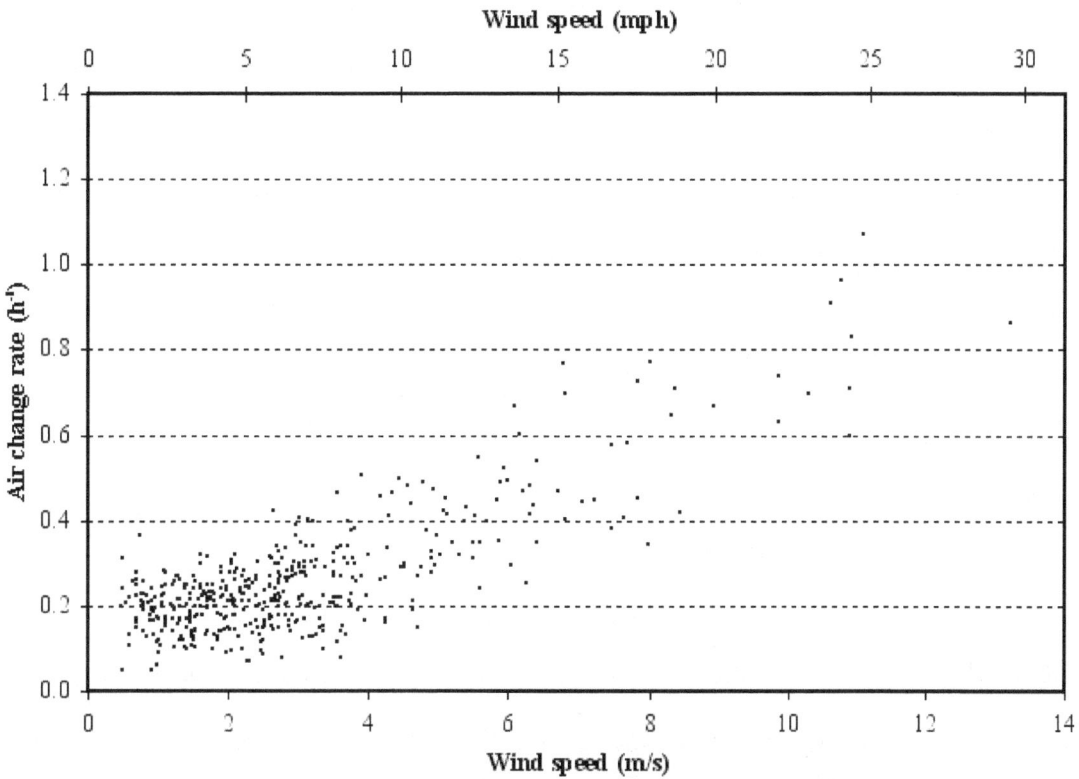

Figure 8: Air change rate versus wind speed, Condition 0

The data in Figures 7 and 8 show a significant fraction of the air change rates under infiltration-only conditions below the 0.25 h^{-1} assumption for infiltration in the HUD MHCSS. For these data, 22 % of the infiltration rates are below this value. Note that this fraction is strongly dependent on the weather conditions existing during the tests and that the tests were not performed under weather conditions that represent the annual climate of the test site. Therefore the fraction below 0.25 h^{-1} is not necessarily representative of what might be expected on an annual basis for this house in this climate.

Figure 9 shows the air change rates under Condition 1a (forced-air fan on continuously, outdoor air intake sealed) plotted against indoor-outdoor temperature difference for low wind speeds (less than 2 m/s (4.5 mph)). The air change rates are fairly constant for ΔT greater than 10 °C (18 °F), but increase for lower temperature differences. Note that with the forced-air system on at positive temperature differences (at low wind speeds), the measured air change rates are actually lower than with the system off as seen in Figure 7. This result might not be expected in this home with its significant level of duct leakage. This issue is discussed in more detail in the section on airflow modeling analysis.

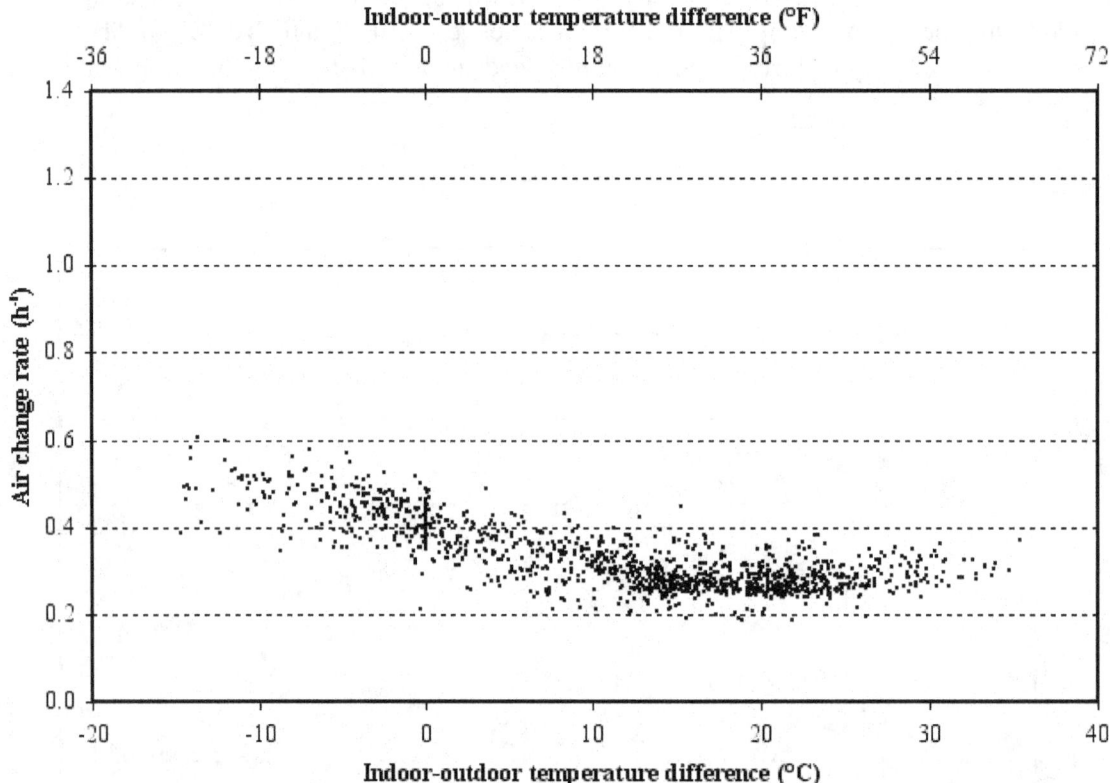

Figure 9: Air change rate versus outdoor-indoor temperature difference, Condition 1a

The dependence of air change rate on temperature difference with the forced air system running continuously and the outdoor air intake open, i.e., Condition 1b, is similar to the dependence under Condition 1a. Though not shown here, the air change rates for Condition 1b tests are slightly higher (about 0.05 h^{-1}) than with the intake closed under Condition 1a. This differential is very close to the measured outdoor air intake flow reported earlier of 7 L/s (16 cfm), which corresponds to 0.08 h^{-1}.

A large number of the air change rates under conditions 1a and 1b are less than 0.35 h^{-1}, the target ventilation rate in the HUD MHCSS. In fact, 43 % of the values are below 0.35 h^{-1} under Condition 1a, and 18 % are below it for 1b. As was noted in the discussion of the measured infiltration rates, the fraction of values below 0.35 h^{-1} is strongly dependent on the weather conditions existing during the tests. Since the tests were not performed under weather conditions that represent the annual climate of the test site, these fractions are not necessarily representative of annual performance. Also, these two conditions only reflect operation with the forced-air fan operating continuously, which is not necessarily consistent with typical operation.

Figure 10 shows the relationship between air change rate and wind speed for ΔT between -10 °C and +10 °C (-18 °F and +18 °F) under test Conditions 1a. The dependence on wind speed is similar to that seen under condition 0 in Figure 8, though the air change rates are higher with the forced-air fan operating.

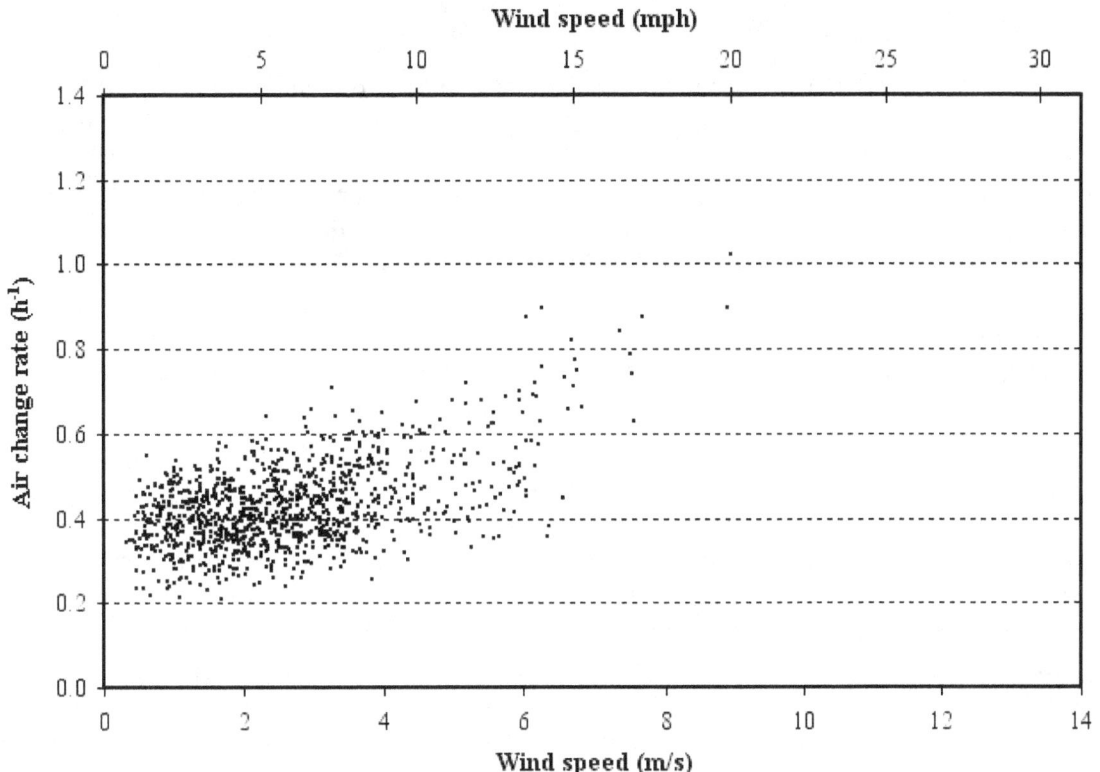

Figure 10: Air change rate versus wind speed, Condition 1a

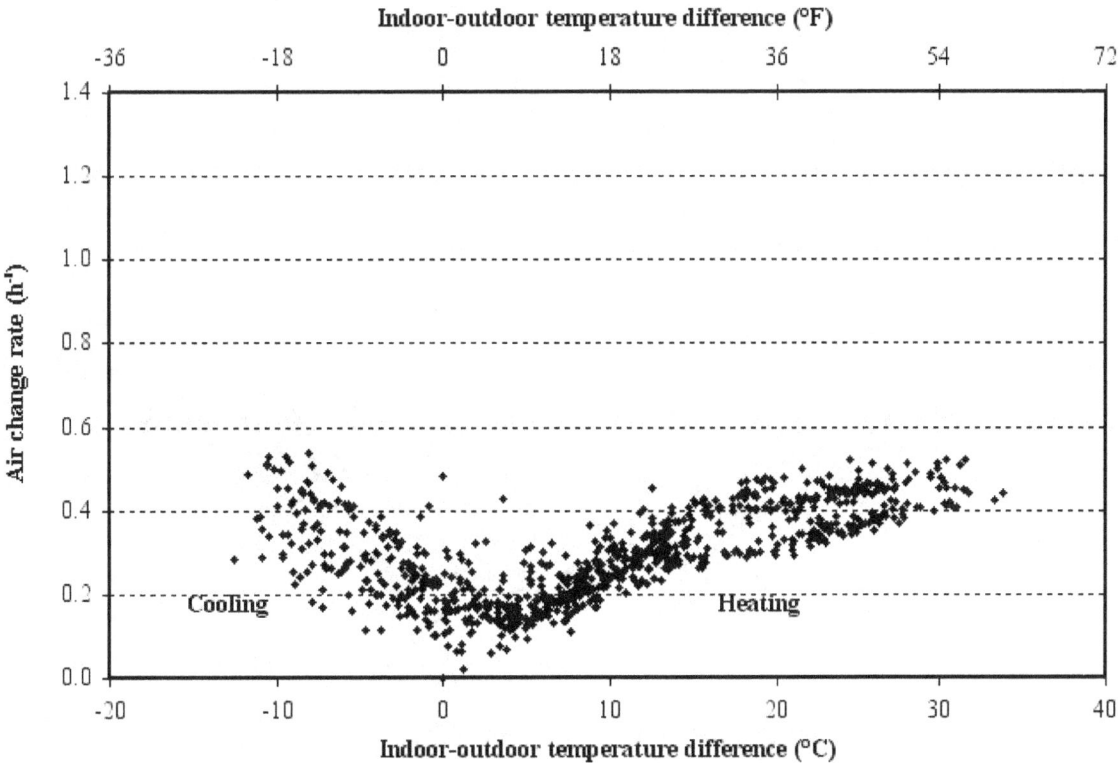

Figure 11: Air change rate versus indoor-outdoor temperature difference, Condition 2a

Figure 11 shows the relationship of the air change rate to indoor-outdoor temperature difference for low wind speeds under test Condition 2a, i.e., the forced air fan controlled by the thermostat and the outdoor air intake sealed. The data are similar to those seen under Condition 0 in Figure 7 due to the relatively low furnace run-times as discussed later. Though not shown, the dependence of air change rate on temperature difference for Condition 2b, under which the outdoor air intake is open, is similar to that seen for Figure 11. For these two cases, 51 % of the values are below the 0.35 h^{-1} target ventilation rate in the HUD MHCSS under 2a, and 32 % are below this value for Condition 2b. Again, these fractions do not necessarily represent annual performance relative to the standard given that the measurements were not necessarily made under weather conditions that represent the annual climatic conditions.

As noted earlier, the dependence of air change rate on temperature difference is fairly similar for conditions 2a and 2b (thermostatic control of forced-air fan) and for condition 0 (no fan operation). This is partly due to the low "fan on" times, which are associated with the sizing of heating and cooling equipment relative to the actual loads. Figure 12 shows the percent of time the forced air fan is on during each hour when the system is under thermostatic control plotted against the indoor-outdoor temperature difference. Note that almost all of the "fan on" times under heating conditions (right side of plot) are under 50 %, with the system generally running less than 33 % of the time. The cooling system (left side of plot) appears to be less oversized, running as much as 80 % of the time under the hottest conditions, but often running much less.

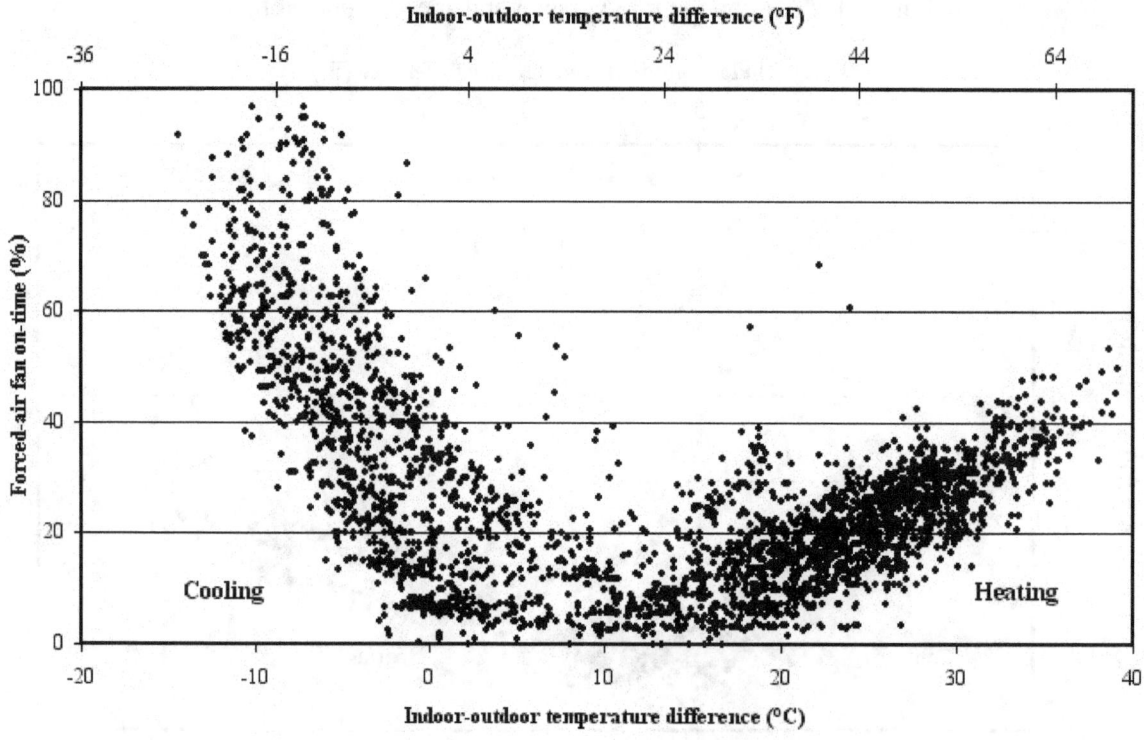

Figure 12: Fan on-time versus outdoor-indoor temperature difference

4.4 Energy Consumption

Pre-retrofit gas (heating) and electric (cooling) energy consumption data was collected over a period of one year from May 2005 to May 2006. Figure 13 is a graph of the energy required to cool the building, with each point corresponding to 24 h of cooling energy consumption, versus the average indoor-outdoor temperature difference during that period. These values are based on the electrical energy consumed by the air-conditioning system and the system specified COP of 2.93. The graph shows the cooling energy consumption when the forced air fan (FAF) is always on, condition 1, and when the fan operation is controlled by the thermostat, condition 2. Comparing the data for the two conditions, one can see the trend of slightly lower energy consumption when the fan is thermostatically controlled.

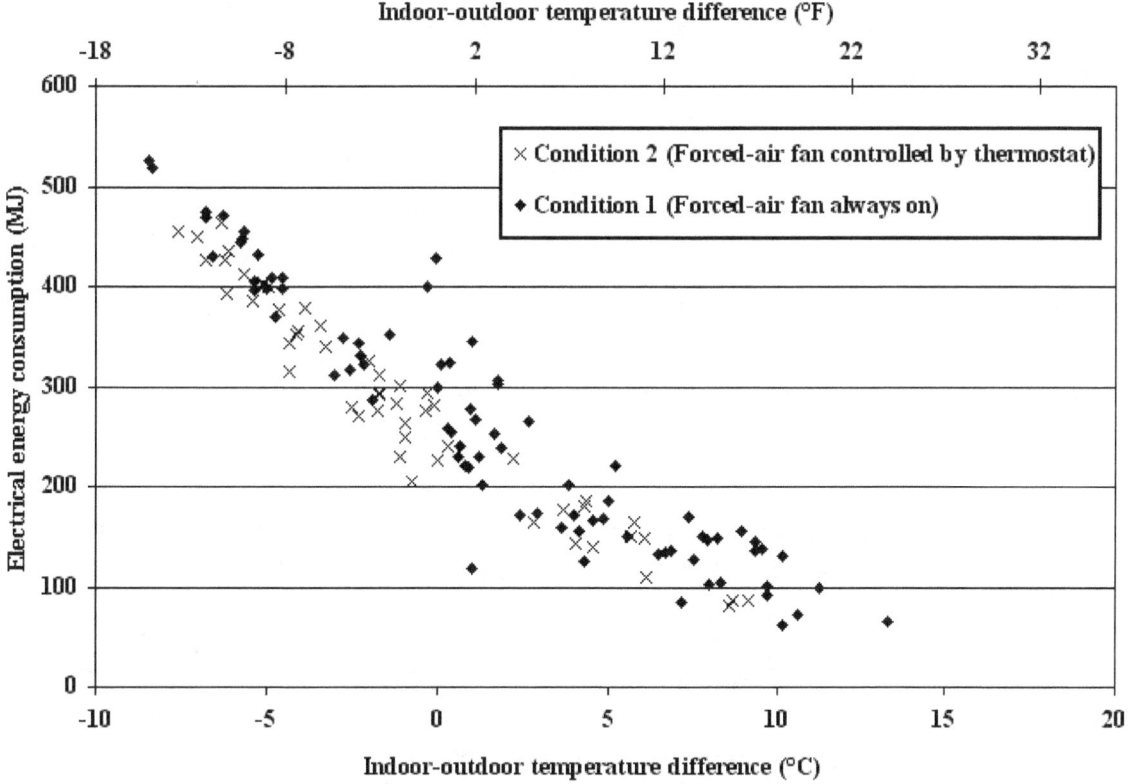

Figure 13: Air conditioning energy consumption versus indoor–outdoor temperature difference

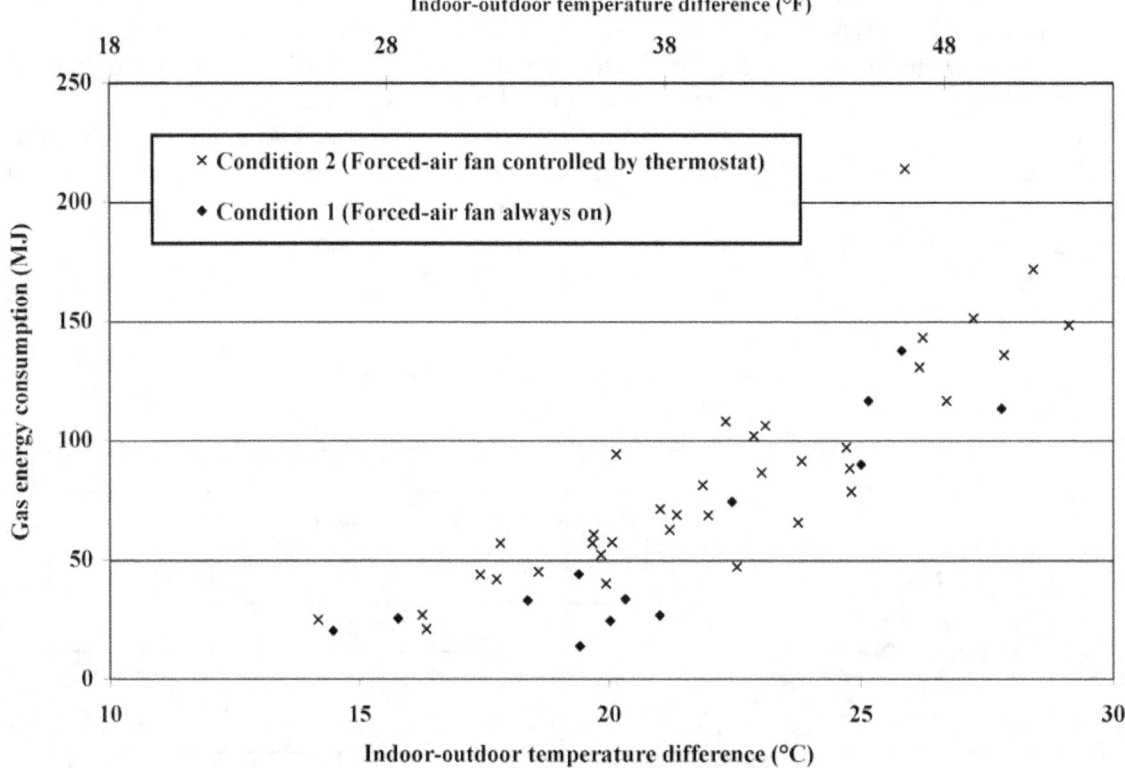

Figure 14: Heating energy consumption versus indoor–outdoor temperature difference

Figure 14 is a graph of the energy used to heat the building, with each point again corresponding to a 24 h interval. These values are based on the rate of gas consumption times the heating system efficiency of 85 %.

Analysis of the data in the two figures yields slopes of about 22 MJ/°C (11600 Btu/°F) for the cooling data in Figure 13 and 10 MJ/°C (4700 Btu/°F) for the heating data in Figure 14. These two values should be the same based on conduction heat losses (gains) alone, but their being within a factor of two of each other is not unreasonable given the use of assumed values for the system efficiencies, the impact of solar and other internal gains, the lack of consideration of latent loads, and the variation of infiltration with weather. A simple heat loss calculation for this building, assuming an air change rate of 0.5 h^{-1}, yields a heat loss (gain) rate through the envelope of 15 MJ/°C (7900 Btu/°F), which is roughly halfway between the two measured values.

5. AIRFLOW MODELING ANALYSIS

In order to explore alternate system configurations and the impacts of airtightness and system retrofits, a model of the house was developed in the multizone airflow and indoor air quality model CONTAM (Walton and Dols 2005). This effort also serves as a model validation exercise to supplement previous efforts (Emmerich 2001). The model contained four levels: crawl space, belly volume, living area and attic. The duct modeling capabilities of CONTAM were employed to model the forced-air system. The leakage values of the airflow paths in the model are listed in Table 5. The leaks in the living space envelope include the exterior wall, the interfaces between the ceiling and wall, the floor and wall, and the walls at the corners. In addition, there are two sizes of windows, exterior doors and the living space floor, which contains openings into the belly. There are also a number of interior airflow paths, including leaks in the walls, door frames and open doors. Note that for all the tests described in this report, all interior doors were open. The attic has leakage in its "floor," i.e., the ceiling of the living space, as well as the two types of vents to the outdoors. The crawl space has leaks to the outdoors in the walls, vents located in the front and rear of the house, and an access door. The model also includes a leak from the crawl space into the belly. Finally, duct leakage into the belly, based on the measurements described earlier, is included in the model.

Table 5: CONTAM model leakage values

	Exterior airflow paths	**ELA at 4 Pa**
Living space envelope	Exterior wall	0.14 cm^2/m^2
	Ceiling wall interface	0.81 cm^2/m
	Floor wall interface	1.24 cm^2/m
	Window #1	5.00 cm^2
	Window #2	1.94 cm^2
	Corner interface	0.808 cm^2/m
	Exterior doors	18.7 cm^2
	Living space floor to "belly" volume	3.65 cm^2/m^2
Interior airflow paths	Interior walls	2 cm^2/m^2
	Bedroom doorframe	410 cm^2
	Open interior doors	2 m x 0.9 m
	Bathroom doorframe	330 cm^2
	Interior doorframe	250 cm^2
	Closet doorframe	4.6 cm^2
Attic	Attic floor	2 cm^2/m^2
	Roof vents	0.135 m^2/each
	Eave vents	106 cm^2/m
Crawl space and belly	Exterior walls of crawl space	25 cm^2/m^2
	Rear crawl space vents	323 cm^2
	Front crawl space vents	465 cm^2
	Crawl space access door	206 cm^2
	Crawl space to "belly"	258 cm^2
	Duct leak into belly	320 cm^2

Figure 15 shows the measured and predicted air change rates with the forced-air system off as a function of indoor-outdoor air temperature difference (ΔT) under low wind speed conditions. The values predicted with the CONTAM model of the house are in good agreement with the measurements, particularly at low values of ΔT. The model tends to under predict by about 20 % at higher values.

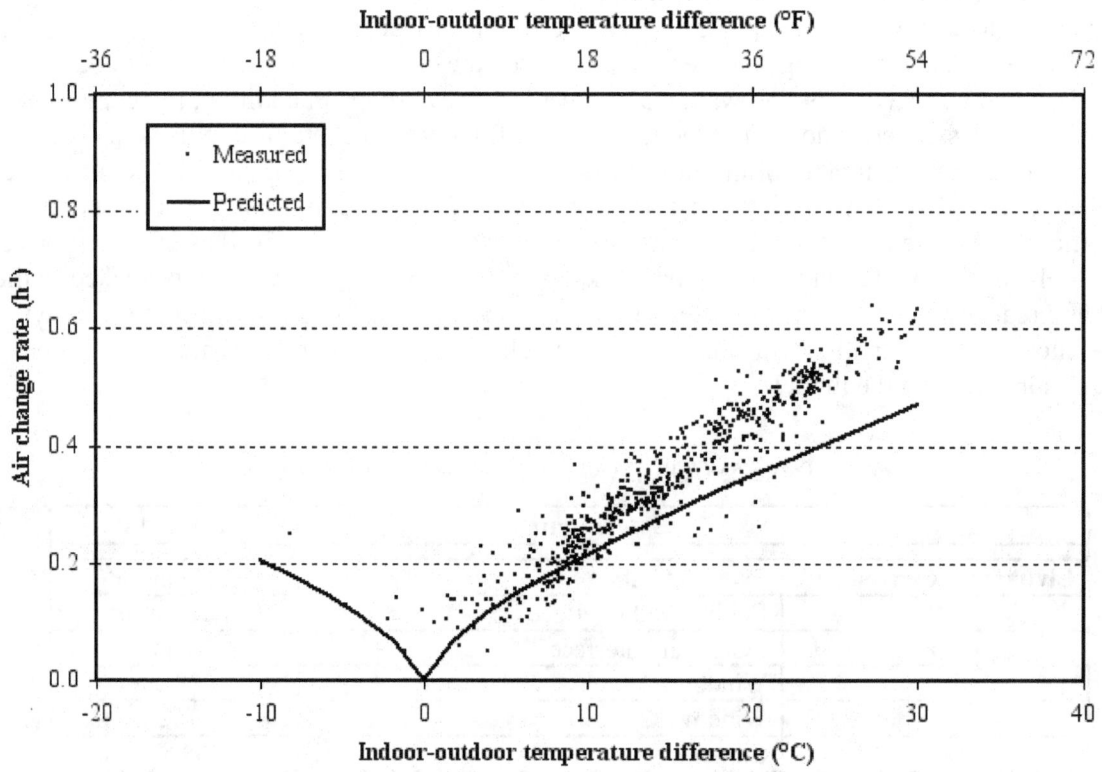

Figure 15: Measured and predicted air change rates versus temperature difference (system off; low wind speed)

Figure 16 is a plot of the measured and predicted air change rates as a function of wind speed for indoor-outdoor temperature differences between -10 °C and +10 °C (-18 °F and +18 °F), again with the HVAC system off. The predicted values in the figure correspond to a temperature difference of 6.0 °C (10.8 °F). The line of predicted values falls in the mid-range of the measured data. Overall, the agreement between the predicted and measured values is quite good.

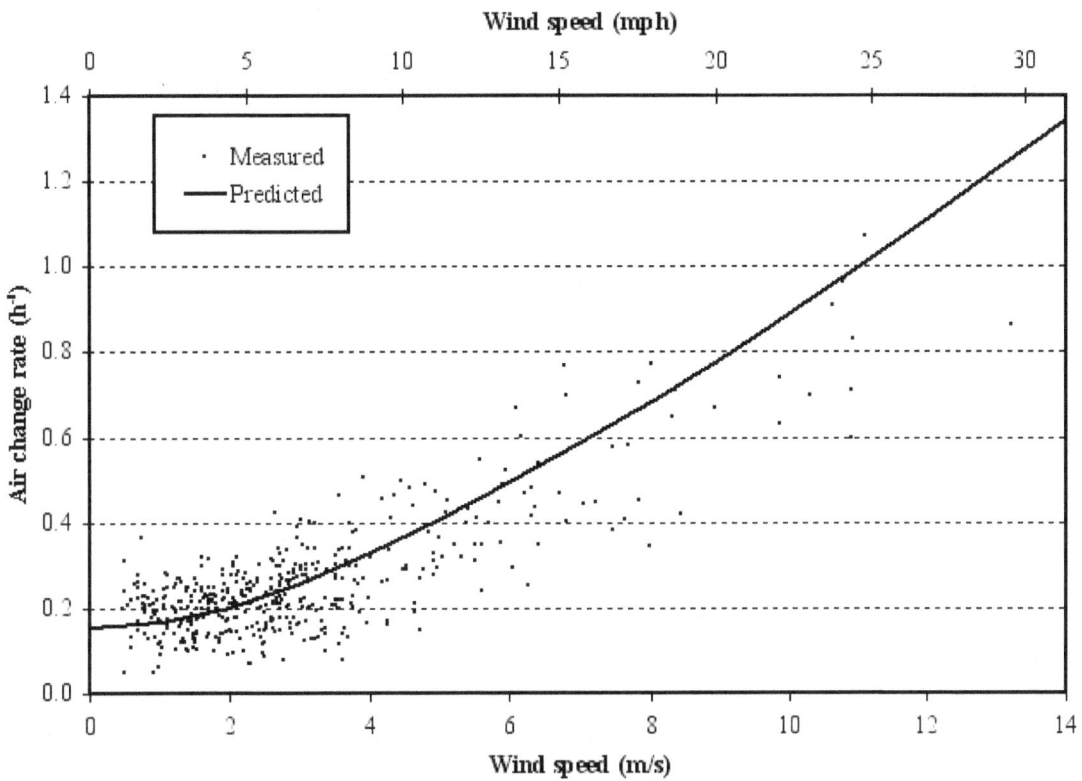

Figure 16: Measured and predicted air change rates versus wind speed
(system off; low temperature difference)

Figure 17 presents the measured and predicted air change rates as a function of temperature difference at low wind speeds with the forced-air fan on. As noted previously, the air change rates under positive temperature differences are actually lower than with the system off, which might not be expected with significant duct leakage. Airflow measurements indicate that the system moves about 470 L/s, but about 120 L/s is lost through duct leakage into the belly. While some of this airflow returns to the living space via leaks in the floor, the remaining air that flows through the crawl space to the outdoors tends to depressurize the house. However, at positive values of ΔT, the duct leakage into the belly "competes" with the stack effect, decreasing the air change rate into the house. This effect has actually been proposed as a means of controlling airflow and contaminant entry from crawl spaces (Phaff and De Gids 1994). Overall, with the fan

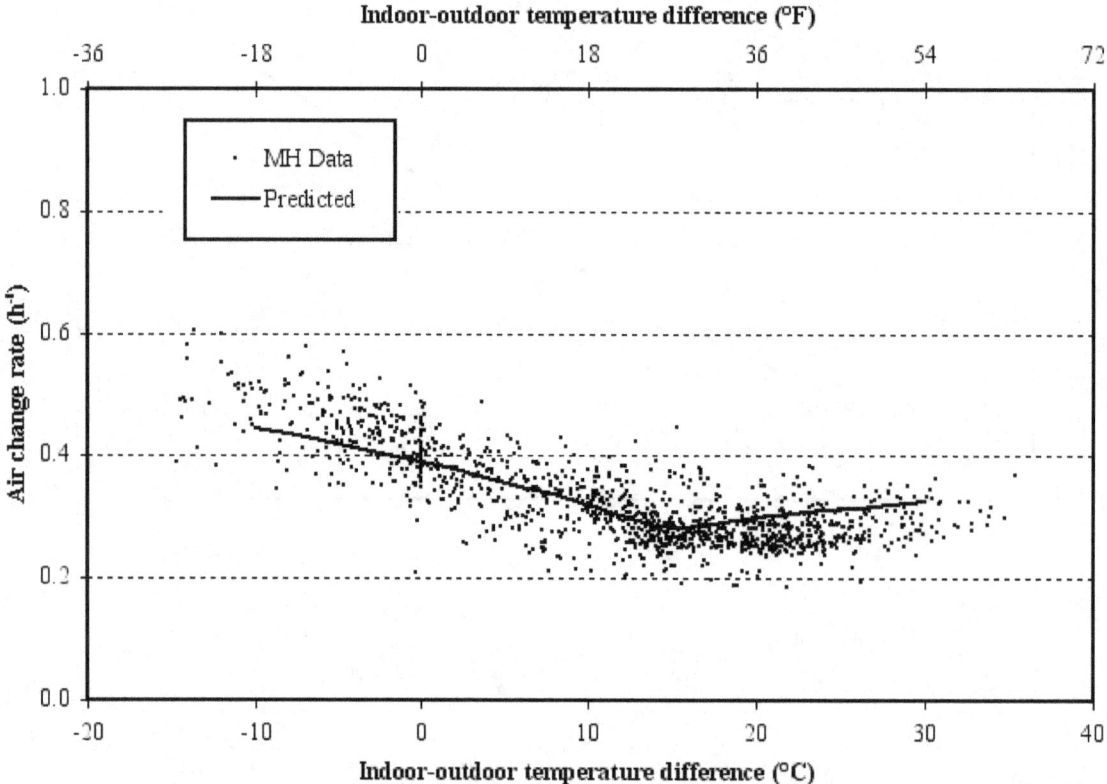

on, the agreement between the predicted and measured air change rates is quite good.

Figure 17: Measured and predicted air change rates versus temperature difference (system on; low wind speed)

Figure 18 presents the measured and predicted air change rates as a function of wind speed with the forced-air fan on. Both the predicted and measured values are fairly independent of wind speed below about 4 m/s (10 mph), but increase approximately linearly for higher wind speeds. The predicted values are on the order of 0.1 h^{-1} lower than the measurements.

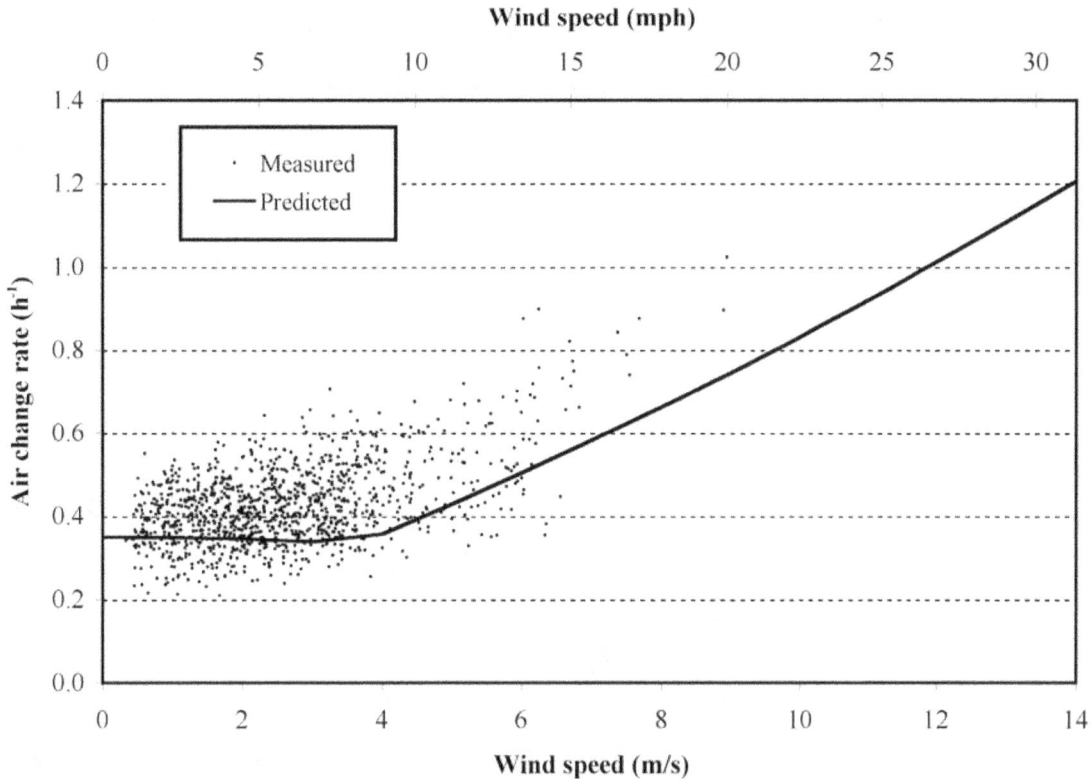

Figure 18: Measured and predicted air change rates versus wind speed
(system on; low temperature difference)

Figure 19 presents the measured and predicted air change rates as a function of temperature difference at low wind speeds with the forced-air fan controlled by the thermostat. As noted earlier, the heating and cooling systems are somewhat oversized and therefore the system does not run for long periods of time except under very cold or very hot weather. Therefore, the air change rates in this figure exhibit a pattern that is similar to that seen in Figure 15 with the fan off. The CONTAM predictions are in good agreement with the measured values.

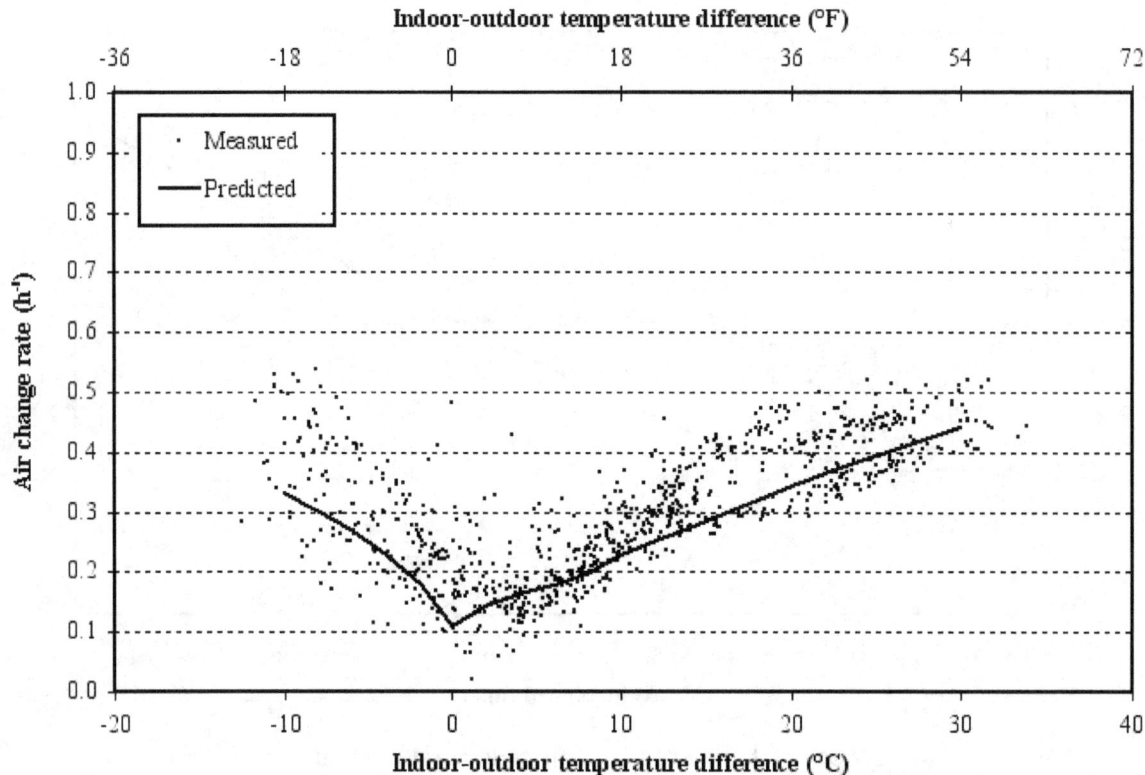

Figure 19: Measured and predicted air change rates versus temperature difference (system controlled by thermostat; low wind speed)

Figure 20 presents the measured and predicted air change rates as a function of wind speed with the forced-air fan controlled by the thermostat. Again the general pattern is similar to that seen with the fan off, and the agreement between measured and predicted is very good.

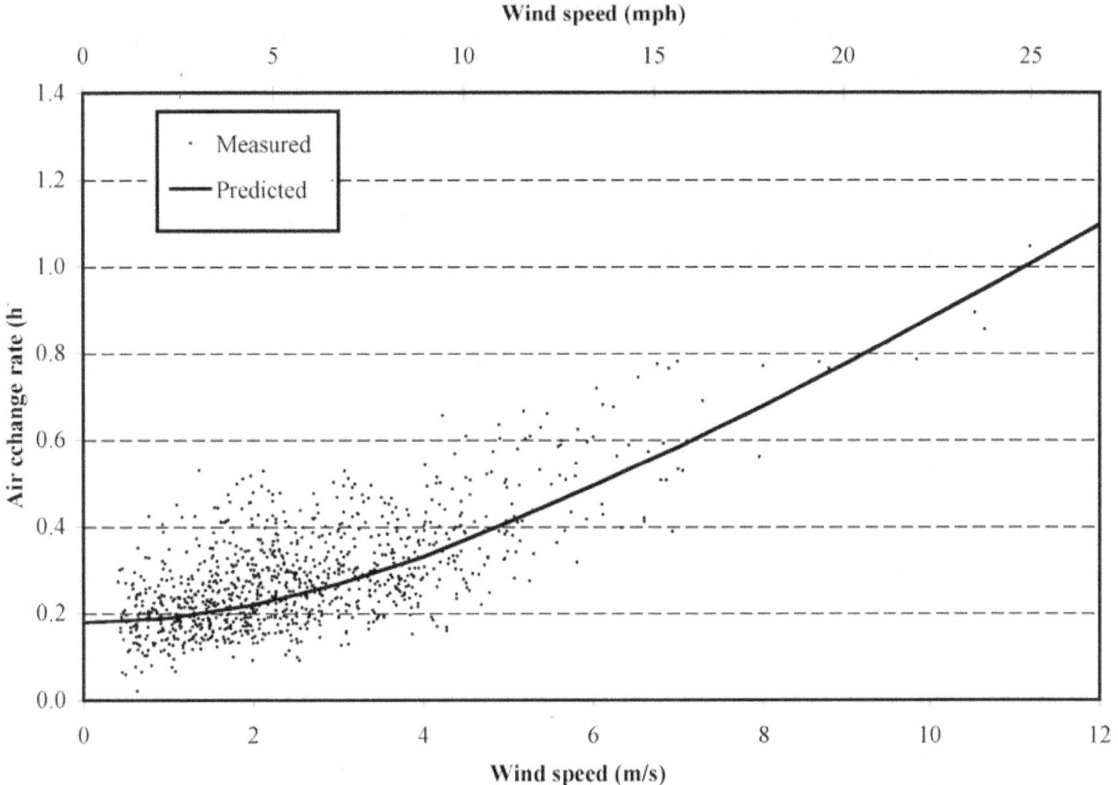

Figure 20: Measured and predicted air change rates versus wind speed
(system controlled by thermostat; low temperature difference)

The CONTAM model of the manufactured house does an excellent job predicting building air change rates as a function of weather conditions and system operation. This model will therefore be useful in later stages of this research project to examine the impacts of building and system retrofits on building ventilation as well as for studying the these building performance issues in different climates.

6. DISCUSSION

This report has provided a characterization of a double-wide manufactured home in terms of measured airtightness, air change rates and energy consumption for heating and cooling. These measurements were made in anticipation of the house being retrofit to increase the airtightness of the envelope and air distribution ductwork, as well as to verify the results of an earlier modeling study (Persily and Martin 2000).

The building airtightness as measured by whole house fan pressurization yielded a value of 11.8 h^{-1} at 50 Pa, which is relatively leaky compared with recently-constructed manufactured houses (Persily and Martin 2000). In addition, the measured duct leakage into the belly of the house is relatively large compared with the field data available. The duct leakage is due in part to some damage to the supply ductwork that occurred during construction and/or installation, which in combination with the overall duct leakage leads to nonuniform supply air distribution to the various rooms of the house. Also, the measured airflows of the bath and kitchen exhaust fans are below their rated values, as well as below the HUD requirements.

The measured air change rates exhibit a dependence on weather conditions and system operation that is consistent with the previous modeling study, but the airflow dynamics in the real building are more complex than the idealized model used in that previous analysis. The infiltration rates, with the system off, do exhibit the expected dependence on temperature difference and wind speed, with many of the values below the 0.25 h^{-1} assumption for infiltration in the HUD MHCSS. Although the measurements were not performed under weather conditions representing the annual climate of the test site, about one-fifth of the measured infiltration rates were below 0.25 h^{-1}, even in this relatively leaky house. The air change rates with the forced-air system running exhibit an unexpected dependence on temperature difference due to the pressure dynamics created by the duct leakage into the belly of the house. These rates, with the outdoor air intake into the system return side, should be in compliance with the HUD requirement of 0.35 h^{-1}, however the measured values are often below this air change rate. Specifically, about 20 % of the measured values fall below this requirement with the forced-air fan operating continuously. With the forced-air fan controlled by the thermostat, perhaps a more realistic operating scenario, about one-third of the measured air change rates are below 0.35 h^{-1}. The modeling of the house with CONTAM revealed the ability to predict the measured air change rates under a range of weather and system operating conditions.

The next phase of the study will involve repeating these measurements after the envelope and duct system tightening retrofits. The envelope retrofits include the installation of a house wrap over the exterior walls and sealing a number of large leakage sites in the floor of the house. The house wrap was installed by removing the exterior siding, installing the house wrap with all joints taped, and covering it with new siding. Leakage sites in the floor of the house were sealed with a polyurethane foam sealant. The areas sealed included the floor portion of the marriage line between the two sections of the house, holes in the floor made for the P-traps in the bathrooms, and other openings associated with utility penetrations. The ends of the two main air distribution ducts were sealed with polyurethane foam, as well as their intersections with the two crossover ducts that connect them to each other. The vertical ducts were also sealed with mastic at both ends where they connected to the main duct and the floor registers. The impact of these retrofits on airtightness, air change rate and energy consumption will be reported in a future publication.

7. ACKNOWLEDGEMENTS

The authors wish to express their appreciation to Robert Zarr of NIST and Michael Lubliner of Washington State University for their many helpful comments in reviewing this document.

8. REFERENCES

ASHRAE 2007. ANSI/ASHRAE Standard 62.2. Ventilation and Acceptable Indoor Air Quality in Low-Rise Residential Buildings. American Society of Heating, Refrigerating and Air-Conditioning Engineers.

ASTM. 2003a. Standard Test Method for Determining Air Leakage Rate by Fan Pressurization, E779-03. American Society for Testing and Materials.

ASTM. 2003b. Standard Test Method for Airflow Calibration of Fan Pressurization Devices, E1258-88(2003). American Society for Testing and Materials.

ASTM, 2003c. Standard Test Methods for Determining External Air Leakage of Air Distribution Systems by Fan Pressurization, E1554-03. American Society for Testing and Materials.

ASTM. 2006. Standard Test Method for Determining Air Change in a Single Zone by Means of a Tracer Gas Dilution, E741-00. American Society for Testing and Materials, Philadelphia, PA.

Hales, D., B. Davis, and R.B. Peeks. 2007. Effect of mastic on duct tightness in energy-efficient manufactured homes. ASHRAE Transactions, 113(2).

Emmerich, S.J. 2001. Validation of multizone IAQ modeling of residential-scale buildings: a review. ASHRAE Transactions, 107(2).

HUD. 1994. Part 3280, Manufactured Home Construction and Safety Standards. U.S. Department of Housing and Urban Development.

Lubliner, M., D.T. Stevens, and B. Davis. 1997. Mechanical Ventilation in HUD-Code Manufactured Housing in the Pacific Northwest. ASHRAE Transactions, 103 (1).

Persily, A.K., Martin, S. 2000. A Modeling Study of Ventilation in Manufactured Houses. NISTIR 6455, National Institute of Standards and Technology.

Phaff, HJC and deGids, WF. 1994. The Air Lock Floor. Proceedings of 15th Air Infiltration and Ventilation Centre Conference.

Walton, G.N. and W.S. Dols, CONTAMW 2.4 User Guide and Program Documentation. 2005. NISTIR 7251. National Institute of Standards and Technology.

www.ingramcontent.com/pod-product-compliance
Lightning Source LLC
Chambersburg PA
CBHW081814170526
45167CB00008B/3434